What Tinkering?

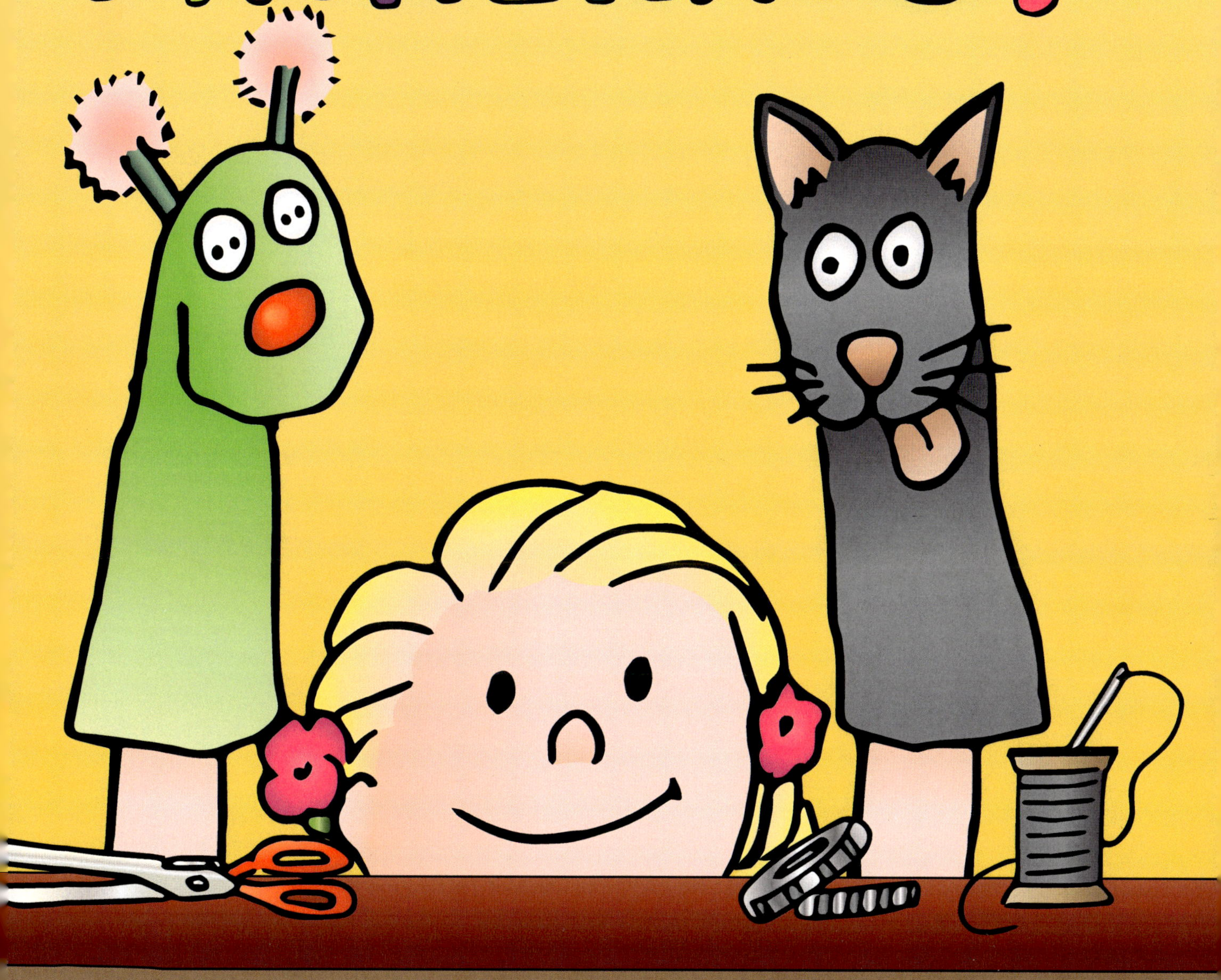

Written by April Smith
Illustrated by Dalila V. Smith

Text copyright © 2024 Performing in Education, LLC

Illustrations copyright © 2024 Dalila V. Smith

All rights reserved. No part of this publication may be reproduced, distributed, or transmitted in any form or by any means, including photocopying, recording, or other electronic or mechanical methods, without the prior written permission of the publisher, except in the case of brief quotations embodied in critical reviews and certain other noncommercial uses permitted by copyright law. For permission requests, write to the publisher at the address below.

Performing in Education LLC
help@performingineducation.com
500 N. Estrella Parkway #B2 #496
Goodyear, AZ 85338

First Edition, 2024

ISBN: 979-8-89217-110-6

Typeset in Gill Sans

Special thanks to Rebecca Riddle, Sarah Wilson, and Lydia Pearson for their contribution to this series.

Visit us at SimplifyScience.com for our companion curriculum.

A Note For Teachers & Home Educators:

After searching for vocabulary-rich picture books for the science standards and coming up empty-handed, we decided to create this our own. We hope you find this book to be a valuable resource. Please share your experience with our science picture books by emailing us at **help@simplifyscience.com**.

View our other titles at **simplifyscience.com/books**.

Find hands-on lessons for the science standards at
simplifyscience.com.

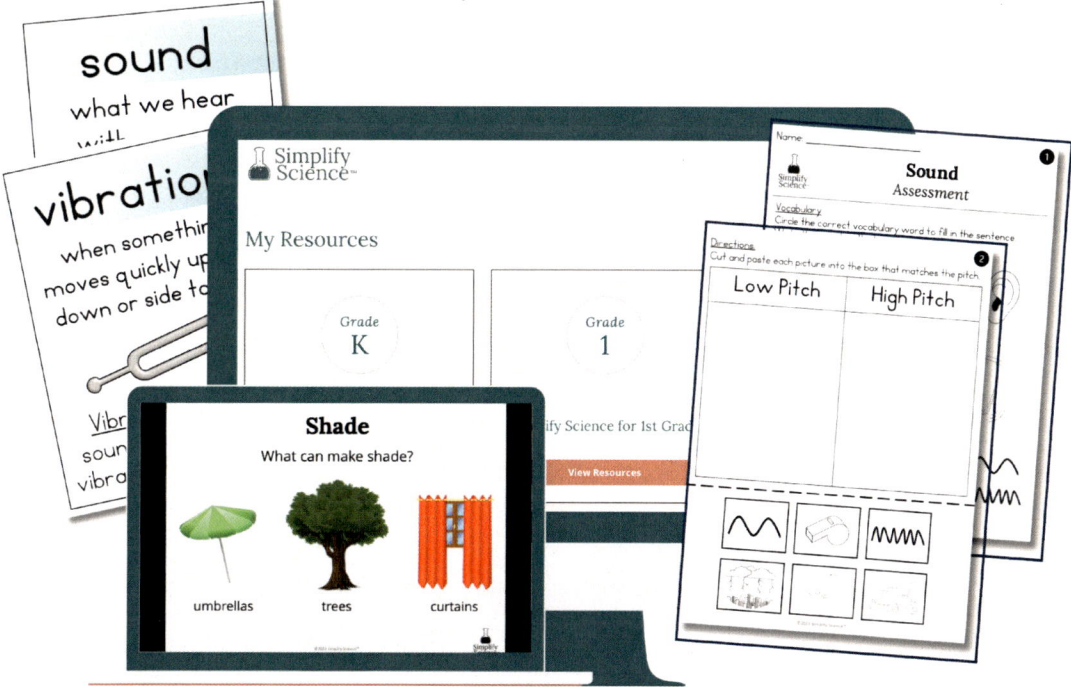

Welcome to the world of tinkering!

Tinkering is all about exploring objects.
Objects are anything you can see or touch.

What objects do you see in this tinkering studio?

Every object has pieces. **Pieces** are small parts that make up an object.

Pieces are put together to assemble objects. When you **assemble** something, you put it together. Have you ever assembled something?

Tinkering involves taking things apart. When you take something apart, you're **disassembling** the object.

What do you see?

Tinkering is discovering how things work.

When Jackson took apart the toy car, he learned that the gears help to control the wheels.

Sometimes, tinkering means making changes.

Before Jackson **reassembled**—or put back together—this toy, he added a special motor that made it super speedy!

Tinkering is about reimagining.

Isla is thinking about how she can repurpose her old stuffed animals. Can you think of any ideas?

When you **repurpose** something, you change the object and give it a new job.

Isla decided to repurpose her stuffed animals by turning them into puppets!

Tinkering is about experimenting.

Dante tried out different ideas before he decided to turn his old night stand into a comfy reading chair!

Tinkering is about trial and error.

If something doesn't work as planned, try another idea! When Luca's tower of cards kept falling, she decided to use a stronger material.

Tinkering is about being creative.

Aaliyah is turning these broken dishes into a work of art.

Pg 24

What characteristics does this creation have? **Characteristics** are special qualities that describe an object, like color, texture, shape, and function.

When you repurpose an item, you change its function. An object's **function** is how it is used. These broken dishes were used to eat off of, but now they are used as a beautiful work of art to enjoy!

Tinkering is about problem-solving.

Kai wants to take her dinosaur toys in the pool, but they don't float. What can she do?

Pg 28

Kai uses materials she found at home to build a boat for her toys.

Now her dinosaurs can float with her in the pool!

Tinkering is like a puzzle!

Sara loves to take apart pens and put them back together.

Quinn enjoys taking apart flashlights and putting them back together.

Pg 32

Tinkering is about endless possibilities!

What will you tinker with next?

Pg 34

Glossary

Assembled: put together

Characteristics: special qualities that describe an object

Disassembled: taken apart

Function: how an object is used

Object: anything you can see or touch

Pieces: small parts that make up an object

Reassembled: put together again

Repurpose: to change an object and give it a new job

Made in United States
Orlando, FL
12 June 2024